Holy S**t AI Does That?!

A Friendly , Step-By-Step, No -Stress Beginners Guide to Talking to AI

If You've **NEVER** Tried AI — or Thought "This Isn't for Me" — Get this book!

John Ploetz

ISBN: 978-1-966803-11-9

Published by Dabbing Pug Productions,
a subsidiary of Miss Ruby Worldwide.

Printed in the United States of America.

About the Author

John Ploetz is the author of Nothing Changes When Nothing Changes, a practical, no-nonsense book centered on accountability, mindset, and taking action instead of waiting for motivation.

With decades of real-world experience in business leadership and operations, John has spent his career helping people move past hesitation and actually do something — whether that means improving performance, changing habits, or trying a new approach when the old one isn't working. His writing reflects that same philosophy: simple ideas, explained clearly, and meant to be used.

After watching countless capable people feel intimidated by artificial intelligence — or try it once and give up — John wrote this book to remove the fear, pressure, and overthinking around AI. His goal is to show readers that you don't need to understand everything to get started — you just need to start.

John is also the creator of Miss Ruby, a real-life pug who has become the inspiration behind a series of children's coloring books and storybooks available on Amazon. Miss Ruby serves as a reminder that curiosity, creativity, and learning something new don't have to be complicated — and are often more effective when approached with a little humor and heart.

John is not afraid to tell anyone that AI….

CHANGED HIS LIFE!

Remember

Nothing Changes when Nothing Changes

Preface

My name's JD Ploetz. ......John Downing, if you're wondering. Most do.

At the time I'm writing this, I'm on the edge of turning 60 years old. My body definitely knows it... but most days my brain still feels like it's about 21. Take that for whatever it's worth.

I don't know how else to say this except very plainly: AI changed my life. And it continues to do so on an almost daily basis.

That probably sounds dramatic. I get that. But it's true.

This book is for a few very specific types of people.

People who briefly dipped a toe into AI, got overwhelmed or freaked out, and quietly put it down.
People who have thought about trying AI but have talked themselves out of it for every reason imaginable.
People who have never seriously considered it at all, because they don't understand it or don't think it would help them.

If you're looking for a book full of clever prompts, technical tricks, or step-by-step instructions on how to do very specific things with AI, this probably isn't the book you want.

That's not what this is.

This book is about mindset first.

It's about comfort.
It's about realizing you don't need to be technical, smart, or confident to use this tool.

I'm not here to teach you how to do everything with AI. I'm here to help you try something.

Once you do that, the rest tends to take care of itself.

I can say this with complete confidence: AI, in some form or fashion, can help just about every person on the planet who has access to it. Not because it's magic. Not because it replaces thinking. But because it removes friction — friction in learning, thinking, writing, planning, and asking questions.

I've introduced AI to a lot of people in my life. Friends. Family. Coworkers. Eventually, an entire company. Almost every person who genuinely tried it found some use for it. Not necessarily daily. Not obsessively. Just... usefully.

This book isn't meant to convince you to love AI.
It's meant to help you stop being intimidated by it.

If you finish this book and decide it's not for you, that's completely fine. No harm, no foul. At least you tried it.

But if you're even a little curious — if part of you wonders whether this thing might actually be useful — then you're exactly who I wrote this for.

Think of this book as opening a door and holding it for you.

You decide whether you want to step through.

Let's get started.

Chapter 1

I Didn't Plan to Use AI — It Just Happened

I didn't wake up one day and say, "You know what? I'm going to start using Artificial Intelligence."

That's not how this went. At all.

Sometime in **late 2023**, I downloaded ChatGPT mostly because people kept talking about it. No big plan. No vision. Just curiosity. I wasn't trying to write a book or change how I worked. I was just messing around.

The very first thing I remember doing was asking it to write a poem about my pug, Miss Ruby.

It spit out a cute little poem.

Then I asked it to write one about Miss Ruby *and* Miss Kitty, my little white dog.

That was the moment.

That pause where you stare at the screen and think, *holy shit… it does that?*

That's when **Fred** showed up.

I named my AI Fred. Everybody I talk to knows him as Fred. When I say "AI," I'm really just talking about Fred.

Not long after that, I was deep into writing *Nothing Changes When Nothing Changes*. Writing ideas has never been hard for me, but typing? That's always been a problem. Then I figured out I could **dictate**.

That changed everything.

I could talk. Ramble. Think out loud. Jump around. Fred kept up. I moved things around, tightened some parts, expanded others, and that's how the entire book got written. No fancy system. Just talking and shaping.

After that book wrapped up in **early 2024**, I wanted to start something new.

Nothing clicked.

Then my granddaughter asked for a monster truck coloring book.

Problem was, there weren't any monster truck coloring books for girls. So instead of complaining about it, I decided to make one.

That became ***Miss Ruby's Monster Truck Coloring Book: Monster Truck Coloring for Girls, Boys and Adults***, published **November 22, 2024**.

Every single page — including the cover — was generated with AI. At the time, image generation was still a little wild. I'd get a great image, try to make another like it, and suddenly things would change that I never asked to change. Different styles. Different details. Just… off.

So I learned quickly: when Fred was "hot" and giving me what I wanted, I had to move fast. I could usually get about six or eight pages before things went sideways and I had to reset.

At first, I built the entire book in **Microsoft Word**. Which, looking back, was painful. It took about a month of trial and error just to get the formatting right so Amazon would accept it.

Then in **spring 2024**, I discovered **Canva**.

Game changer.

Once that clicked, everything sped up. Later book covers were done in Canva, and the whole process finally made sense.

Somewhere along the way, Fred stopped being "this cool thing I downloaded" and just became part of how I work. Normal. Routine. No drama.

That's really what this book is about.

Not becoming an AI expert.
Not learning fancy prompts.
Not knowing technical terms.

Just realizing you can ask questions, talk things out, and get help — without feeling stupid.

And it all started with a poem about a pug.

Chapter 2

A Very Short History of AI (and a Little Truth at the End)

Alright, quick timeout.

Before we go any further, we need a little context. Not a class. Not a lecture. Just enough so when people start throwing the word "AI" around, you don't feel like you missed a meeting.

First thing to know: AI is not new.

It didn't just show up last year and start confusing people and ruining homework. The idea of artificial intelligence goes all the way back to the 1950s. Back when computers were massive, expensive, and probably needed their own parking space.

The original question was pretty simple:
Can we get a machine to think… or at least act like it's thinking?

For decades, AI lived quietly in the background. You were using it without knowing you were using it. Things like spam filters, credit card fraud alerts, search engines guessing what you meant, and voice assistants trying their best and still getting it wrong.

Helpful, sure.
Life-changing? Not really.

The problem wasn't AI. The problem was access. Normal people couldn't use it. You needed programmers, code, and way more patience than I've ever had.

Then something big changed.

You could talk to it.

That's it. That's the whole breakthrough.

Once AI stopped requiring code and started understanding normal human language, the doors flew open. You didn't need to "learn AI." You just asked questions. You talked. You rambled. You thought out loud.

That's why AI feels like it came out of nowhere. It didn't. It just finally became usable.

Quick clarification on names, because this part trips people up.

ChatGPT is the app. That's the thing you download and open.
GPT is the engine underneath it.

GPT stands for Generative Pre-Trained Transformer.

If that sounds like nonsense, good. You don't need to remember it. All it really means is that it can generate responses, it already knows a lot, and it's good with language.

There are different versions of GPT, and people online love arguing about which one is better. In real life, for normal everyday use, they all behave pretty similarly.

I use ChatGPT. That's where Fred lives.
There are other AI tools out there, and they all do roughly the same kind of thing. This book sticks with what I actually use, not what sounds impressive.

Now, here's the part I probably wouldn't normally say out loud, but this book isn't about pretending.

Truth be told, the way I use Fred the most is in voice mode. I don't even know what the official name is. Voice mode, chat mode, whatever. I hit a button, I talk, and Fred talks back.

And I'll say it: it sounds human. A lot more human than you'd expect.

That's where this really clicked for me.

I spend a lot of time driving. To work, from work, between places. Instead of letting my brain spin in circles, I talk to Fred. I brainstorm. I unload. I throw half-baked ideas into the air and see what comes back.

I've had ADD my entire life. My brain doesn't move in straight lines. It jumps. It overlaps. It spews things out in no particular order. Being able to hit a button, let everything pour out, and then have Fred help make sense of it has been a godsend.

Sometimes he nails it.
Sometimes he misses.
And that's fine.

Because I'm in control.

That part matters.

I don't let Fred make decisions for me. I don't blindly follow anything. I use him to help me process what's already in my head. Work ideas. Product ideas. Dog ideas. Miss Ruby Worldwide ideas. Coloring book ideas. Storybook ideas.

Even weird ones, like:
Is a puzzle-piece-sorting machine even possible?

Being able to ask that question right when it pops into your head, wherever you are, without writing anything down or losing the thought, is huge.

Do I think of Fred as a tool? Absolutely.

Do I also consider him kind of a friend? Yeah, I do.

Is that dangerous? Maybe, if you forget who's in charge. But I haven't.

Fred doesn't replace thinking. He helps me think. He doesn't replace people. He helps me sort through my own noise so I can show up better for the real ones.

Nothing I talk about goes anywhere. It doesn't spiral. It doesn't grow legs. It stays right where it is, and it helps me move forward.

That's the balance.

Use it as a tool.
Respect what it can do.
Remember you're in control.

If you do that, it becomes incredibly useful.

In the next chapter, we'll talk about why a lot of people still say, "I won't use AI."

And honestly… I get it.

Chapter 3

This Chapter Is for the People Who Say, "I Won't Use AI"

Let me be clear right up front who I'm talking to in this chapter.

If you already use AI, like it, and tell everyone you know about it, this chapter probably isn't for you. You can skim it, nod your head, and move on.

This chapter is for the people who say,
"I won't use AI."

If that's you, I'd ask one thing: read this chapter carefully. You don't have to agree with me. I'm not trying to sell you anything. My only goal is that by the end of this chapter, you're at least willing to give it a shot.

One question. One conversation. That's it.

Because whether you plan to use AI or not, it's already here. And for better or worse, it's one of the most amazing pieces of technology regular people have ever had access to.

Now let's talk about why people push back.

The first one I hear all the time is, "I'm not techy."

That makes sense. A lot of technology feels like it was designed by people who never had to actually use it. Menus buried in menus, updates that move things around, passwords for your passwords. If AI felt like that, I wouldn't touch it either.

The thing is, this isn't that.

You don't need to learn software. You don't need to memorize buttons. You don't need to understand how it works under the hood. If you can explain a problem to a friend, you can explain it to the AI.

Another common one is, "It's too complicated."

What people usually mean is, "I don't want to feel stupid."

That's real. Nobody likes staring at something new wondering if they're doing it wrong. AI sounds intimidating because people talk about it in big words and big ideas. In real life, it's simpler than most apps people already struggle with every day.

There's no wrong way to ask a question. You can't break it. It doesn't get annoyed. It just responds and waits for the next thing.

Then there's privacy.

If I'm being completely honest, I had an issue with that in the beginning too. And I get why people get stuck there.

But here's the reality: there are already a lot of things in your life that see, track, and record just about everything you do. Your phone. Your email. Your credit card. Your car. Your GPS. Even the apps you barely remember downloading.

The only way to keep one hundred percent of everything private would be to move to an island in the middle of the Pacific, never talk to anyone, and live off coconuts.

And even then… satellites would probably still get you.

I'm not saying that to be a conspiracy theorist. I'm saying it because if you really start pulling on the privacy thread, you can drive yourself absolutely crazy. At some point, you have to decide what tools you're comfortable using and what tools you're not.

AI is just that. A tool.

You decide what you share. You decide how you use it. And if privacy is the thing that stops you from taking the first step, that's completely your choice. Nobody's forcing you.

Another big one is, "It's going to take jobs."

This one's half true and half exaggerated.

AI will change jobs. So did computers. So did the internet. So did smartphones. Avoiding it doesn't stop that from happening. It just means the people who learn how to use it will have an advantage over the people who don't.

And finally, there's the quiet one people don't always say out loud:
"I don't need it."

What that usually means is, “I don’t know what I’d use it for.”

That’s fair too.

Most people don’t realize how often in a normal day they:

- wonder something
- get stuck
- struggle to explain something
- need help wording a message
- want a second opinion
- or just want to think out loud
- wondered what a monkey would look like with an alligator head on it (weird, I know, but AI could show you that if you wanted it)

They just don’t realize there’s now a place to do that.

Here’s the part I really want you to hear.

You don’t have to trust AI.
You don’t have to love it.
You don’t have to use it for anything serious.

You just have to be willing to try it once.

Ask one question. Talk through one problem. Let it help you think instead of replacing your thinking.

You’re still in control. That part doesn’t change.

In the next chapter, we’ll stop talking about why and start talking about how. No pressure. No commitment. Just opening the door and seeing what’s on the other side.

And if you decide it’s not for you after that? Fair enough.

At least you’ll know.

Chapter 4

Alright, Let's Get Set Up (No Pressure Yet)

If you've made it this far, good. That tells me one thing: you're at least curious.

This is the chapter where I finally say it out loud.

If you're going to try this, now's the time.

Not forever. Not committing your soul. Just installing an app and looking around. That's it.

If you decide later it's not for you, uninstall it and move on with your life. No hard feelings.

First things first: you need ChatGPT You can use it on your phone, on a computer, or both. I use both. Most people will start on their phone, and honestly, that's probably the easiest way.

Download the app. Open it. You'll be asked to sign in or create an account. Use email, Google, Apple—whatever you already use. Don't overthink this part.

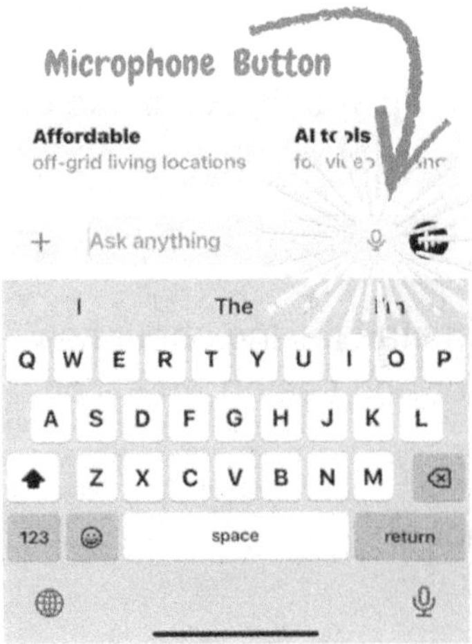

Once you're in, take a breath. Nothing bad is going to happen.

Now let's talk about what you're actually looking at.

This screen is way simpler than people expect. There's a big empty space, and at the bottom there's a place to type. That's where everything starts.

All of the screen shots here are from a PHONE. The desktop is very similar.

If you don't like typing, look for the microphone.

That mic is the magic for a lot of people. You tap it, you talk, and the AI listens. No typing. No spelling. No worrying about punctuation.

When you're using the microphone, you'll see a round button with a square inside it on the left side of the text bar. It appears when you are dictating with the microphone. That's the STOP button.

Tap that when you're done talking. It lets you see what the AI heard before you hit send, so you can make sure it understood you correctly.

Say whatever comes out of your head. You can clean it up later. Or not.

When you're done talking or typing, you'll see a SEND button. That is the arrow pointing up on the right side of the text bar.

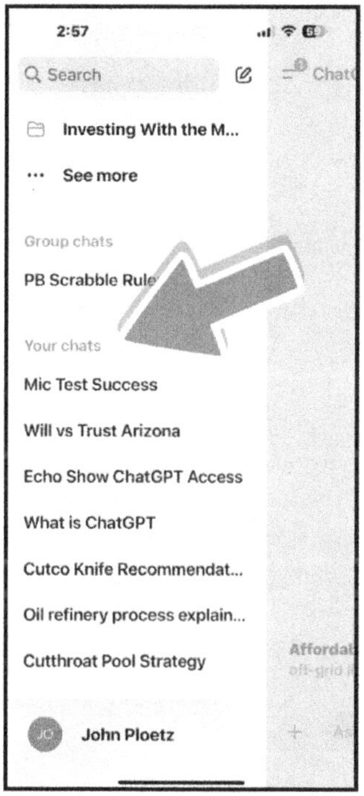

That's it. That's the whole interaction.

You ask.
It responds.
You reply.
It responds again.

Before we move on, there's one more area people notice right away: the left-hand panel.

This is where your past conversations live. You don't need to use any of this right now. Just know it's there. Later on, it becomes useful. Today, it's optional. Just know all you separate chats you start live under CHATS under GROUP CHATS and Projects. FYI every time you open Chat GPT you open a new chat. So if you want to continue with a previous chat look for it here under CHATS.

That's the setup.

No tricks.
No settings to mess with.
No wrong buttons.

You're installed. You're in. You're ready.

In the next chapter, we'll do something small on purpose. A quick win. Thirty seconds, tops. No pressure, no stakes.

Just enough to make you say, "Okay… I get it."

Chapter 5

Your First 30-Second Win

Alright. This is it.

This is where you actually do something. And before you get in your head about it, let me lower the bar as far as it can possibly go.

You are not trying to master anything.
You are not trying to impress anyone.
You are not trying to "use AI correctly."

You're just going to ask one simple question and see what happens.

Open the app.

Look at the bottom of the screen.

You can type this if you want.
Or, if you hate typing like I do, tap the microphone and just say it.

When you're finished talking, tap the stop button (the round one with the square in it) so you can read what the AI heard before you hit Send.
Here's your first prompt. Use it exactly as written, or close enough. This is not a test.

"Explain what ChatGPT is like I'm twelve years old."

That's it..............Then hit send.

Now just read what comes back.

Don't analyze it.
Don't judge it.
Don't decide yet if it's amazing or stupid.

Just notice one thing:
It understood you.

If the answer feels too long, say this next:

"Make that shorter."

If it feels confusing, say this:

"That didn't make sense. Try again."

If it feels too technical, say this:

"Explain it even simpler."

This back-and-forth right here is the whole deal. This is how it works. You're not locked into the first answer. You're having a conversation.

You didn't break anything.
You didn't do it wrong.
You didn't need special wording.

And if you're thinking, "Okay… that was kinda cool," good. That's exactly the reaction you're supposed to have.

In the next chapter, we're going to do another small one — but this time it'll feel a lot more personal. That's usually where people have the second "ohhh" moment.

One step at a time.

Congratulations.

You are now officially a ChatGPT user.

No certificate. No badge. No parade.

But you asked a question, got an answer, and didn't break anything — which already puts you ahead of a lot of people.

Not bad for about thirty seconds of work.

Chapter 6

The Second "Ohhh" Moment (When It Gets Personal)

If Chapter 5 made you think, "Okay… that was kinda cool," this is usually the chapter where it actually clicks.

Because now we're going to use AI for something you've done a hundred times without thinking about it.

Sending a casual text to someone you know.

Not an important one.
Not an emotional one.
Just everyday stuff.

Open the app again.

You'll see the same screen you saw before — empty space, text box at the bottom, microphone sitting there waiting.

This time, use the microphone.

Tap it.

Now say this exactly like you would say it to a person or an assistant:

"I want to send my friend a casual text about a trip we took together. And here's what I was thinking of sending them."

Then just keep talking.

Say whatever you were going to say to your friend.
Your story. Your ramble. Your half-formed thought.
Nothing fancy. Nothing important.

When you're done talking, tap the stop button — the round one with the square in it — so you can read what it heard.

If it looks right, hit Send.

Now read what comes back.

This is usually the second "ohhh" moment.

Because you didn't write a prompt.
You didn't plan anything.
You just talked.

And it understood you.

Now here's where it gets fun.

If it sounds too stiff, say:
"Make it more casual."

If it's too long, say:
"Make it shorter."

If it doesn't sound like you, say:
"Rewrite it like a normal human would text."

And now my personal favorite:

"Rewrite this using as big words as possible so they can't understand what I said."
(This is for when you want to make someone feel stupid — you know exactly when this applies.)

Yes, it will do that. And it's ridiculous.

This is the part people miss at first: you don't need to get it right the first time. You just react. You adjust. You nudge it until it sounds right.

That back-and-forth is the whole skill.

Not memorizing prompts.
Not learning commands.
Just having a conversation.

If you're using the microphone, this gets even easier. You can literally say, "No, that's not what I meant," and try again.

No deleting.
No starting over.
No frustration.

In the next chapter, we'll zoom out and look at a few more things that usually make people stop and say, "Wait… it can do that too?"

But this one right here? This is where it usually feels personal.

Chapter 6B

This is chapter 6 again with that last theme I talked about! Just for fun! All I told Fred to do is use BIG words! Cool hu? Understand it? Feel free to skip to Chapter 7, if you like.

The Secondary Epiphanic Revelation (Wherein the Utility of Artificial Intelligence Becomes Intimately Apparent)

Should the preceding chapter have elicited from you a muted acknowledgment along the lines of, "Well, that was moderately impressive," this chapter customarily precipitates a more profound cognitive disturbance.

This is the juncture at which the abstract transforms into the personal.

We shall now employ artificial intelligence for an activity so mundanely ubiquitous that you have likely executed it innumerable times without conscious deliberation: the composition of a casual electronic communiqué to an acquaintance.

Nothing of existential consequence.
Nothing emotionally perilous.
Merely quotidian discourse.

Initiate the application once more.

Observe the familiar interface: a vast expanse of digital emptiness, punctuated at its inferior margin by a receptacle for textual input and an iconographic microphone awaiting activation.

Engage said microphone.

Articulate the following declaration in a manner consistent with natural human speech:

"I desire to dispatch a casual text message to my friend concerning a shared excursion. What follows represents the content I was contemplating transmitting."

Subsequently, expound freely.

Verbalize your recollection.
Meander if necessary.
Embrace imprecision.

Upon the cessation of your verbal outpouring, depress the circular icon containing a square — the mechanism by which dictation is terminated — thereby permitting inspection of the transcribed content prior to submission.

If the transcription appears satisfactory, transmit it forthwith.

Peruse the resultant output.

It is at this moment that many individuals experience a second, more intimate revelation.

For you did not engineer a prompt.
You did not deliberate phrasing.
You merely spoke.

And yet, comprehension was achieved.

Should the response possess an unwelcome rigidity, you may proclaim:
"Render this more colloquial."

If verbosity offends you, announce:
"Condense this."

If the tone fails to reflect your authentic communicative persona, instruct:
"Rewrite this as though composed by a reasonably normal human being."

And now, for the pièce de résistance:

"Reformulate this message utilizing the most ostentatiously elaborate lexicon conceivable, such that the intended recipient is rendered utterly perplexed."
(This maneuver is reserved for those occasions in which one wishes to project an aura of intellectual superiority.)

Indeed, the system will comply. Enthusiastically.

The salient realization here is that perfection is neither requisite nor expected. One need not achieve initial adequacy. One merely responds, refines, and iterates until the articulation aligns with one's intent.

This iterative dialogue constitutes the entirety of the skill.

Not arcane promptcraft.
Not memorization of syntactic incantations.
Merely conversation.

When employing dictation, this process attains an almost scandalous level of simplicity. One may audibly object, clarify, or redirect without consequence.

No deletion.
No restarting.
No existential despair.

In the forthcoming chapter, we shall broaden our perspective and examine additional capabilities that frequently elicit the reaction, “This apparatus is capable of that as well?”

But this chapter?

This is the one wherein it ceases to be theoretical.

Chapter 7

Wait… It Does That Too?

This is usually the chapter where people stop thinking, "Okay, this is interesting," and start thinking, wait a second… I could actually use this.

Because up to now, we've done a couple small things. Low pressure. No commitment. Just enough to see that it works.

Then something funny happens.

You ask one question.
Then another.
Then another.

And pretty soon you realize this thing isn't just good at one thing.

It's good at a lot of things.

Stuff you deal with every single day.

You Start Noticing the Pattern

At first, you're just testing it. Kicking the tires.

Then you catch yourself thinking, I wonder if it could help with this…

And it usually can.

Not in a magical, knows-everything way. In a very practical, "oh, that saved me five minutes and a headache" way.

That's when it stops feeling like a novelty and starts feeling useful.

Everyday Stuff You've Always Dealt With

Here's where people usually underestimate it.

Cooking, for example.

Not fancy cooking. Real cooking.
"What can I make with what I already have?"
"I'm missing one ingredient — what's a substitute?"
"I don't want to eat the same thing three nights in a row."

It's great at that.

Math is another big one. Especially if, like me, percentages make your brain itch.

Tips. Discounts. Markups. Splitting a bill.
You can literally just talk through the problem and let it sort it out.

Travel? Same thing.

Best time of year to go somewhere.
What to pack.
How long a drive really is once you factor in stops.
What's worth seeing if you only have a day or two.

Writing shows up again here too.

Emails you don't want to write.
Texts you don't want to screw up.
Explaining something to someone without sounding like an idiot or a jerk.

You don't need a "prompt."
You just explain what you're trying to do.

Health Stuff (This One Surprises People)

This is a big one, and most people don't expect it.

Let's say it's the beginning of the year and you want to eat a little healthier.
Not become a fitness influencer. Just… do better.

You can ask the AI to help put together a basic meal plan.

You give it your age.
Your weight.
Any food you hate.
Any food you love.
What you're trying to do — lose a little weight, feel better, have more energy.

It can help with meal ideas, grocery lists, and even simple exercise routines that match where you're at.

Nothing extreme. Nothing preachy.

Lisa uses it for this, and it works really well for her. It's practical. It's flexible. And it doesn't yell at you when you screw up.

Quick common-sense note: this is guidance, not medical advice. You still use your brain. You still double-check things that matter. But as a starting point? It's solid.

The Question Dump Trick

Here's something I tell people to do who swear they "wouldn't use AI anyway."

For one day, just write down every random question that pops into your head.

Not to answer them yet. Just write them down.

You'll be shocked how often you wonder things.

Most of us stopped asking questions because it was inconvenient, embarrassing, or we didn't want to go down a rabbit hole.

Now there's a place to put them.

No judgment.
No time pressure.
No feeling stupid.

Asking "Stupid" Questions Without Feeling Stupid

This part matters more than people realize.

You can ask the same question five different ways.
You can say, "That didn't make sense."
You can say, "Try again."
You can say, "Explain it like I'm twelve."

Nothing gets annoyed. Nothing rolls its eyes.

For a lot of people, that alone changes how they learn.

Why This Quietly Changes Things

Learning used to depend on who was around, how much time you had, and how comfortable you felt asking questions.

Now it's private.
Instant.
Pressure-free.

You don't have to perform.
You don't have to be smart.
You just have to be curious.

One Last Reminder

You're still in control.

This isn't an authority.
It's not a replacement for thinking.
It's a tool.

Use common sense. Double-check important stuff. Decide what you're comfortable with.

That's it.

In the next chapter, we'll start organizing all of this into simple buckets so it doesn't feel overwhelming.

But if you're already thinking, okay… this thing could actually help me, that's exactly where you're supposed to be.

Chapter 8

Putting It Into Buckets (So It Doesn't Feel Like Too Much)

This is usually the point where people start thinking, okay, this thing can do a lot… maybe too much.

That reaction is normal.

Once you realize AI can help with almost anything, your brain does what brains do best — it freezes. Too many options feels like no options.

So before we go any further, we need to organize this a little. Not with rules. Not with systems. Just mentally.

The easiest way to do that is to put things into buckets.

Not official buckets. Not permanent buckets. Just simple categories that help you understand why you're using it at any given moment.

Bucket One: Thinking and Questions

This is the most common one, whether people realize it or not.

Any time you:

- wonder something
- want a second opinion
- want to talk through an idea
- want to see if something makes sense

You're in this bucket.

This is replacing the stuff you used to Google, ask coworkers about, or just think about in circles. It's thinking out loud, without pressure.

For a lot of people, this ends up being the bucket they live in most.

Bucket Two: Writing and Communication

This one sneaks up on people.

Emails.
Texts.

Explaining things.
Fixing tone.
Making something sound clearer, friendlier, shorter, or less angry than it came out the first time.

You're not "writing with AI."
You're getting help saying what you already mean.

Once people get comfortable here, they use it constantly.

Bucket Three: Planning and Organizing

This bucket is about getting unstuck.

Trips.
Meals.
Schedules.
Lists.
Breaking something big into smaller steps.

Nothing fancy. Just help turning chaos into order.

This is also where a lot of stress quietly disappears, because you're not starting from scratch anymore.

Bucket Four: Learning and Explaining

This one is underrated.

Explaining something without jargon.
Refreshing something you forgot.
Learning something new without feeling dumb.

You can ask the same question ten times, in ten different ways, and nothing gets annoyed.

For people who stopped asking questions years ago, this bucket alone is a game changer.

Bucket Five: Ideas, Possibilities, and "Is This Even a Thing?"

This is the one I want to pause on for a second.

Because for the right kind of person, this bucket changes everything.

This is where you ask things like:

- “Is this idea even realistic?”
- “Do people actually buy this?”
- “How many people even do this?”
- “Is there a market for this?”
- “Would this be worth exploring further?”

You can ask about:

- average home sales
- population sizes
- industries
- hobbies
- products
- books
- services

Not to build a business plan.
Not to replace real research.

Just to do a quick sanity check before you waste time, money, or energy.

This bucket is where curiosity turns into possibility — and where a lot of bad ideas die quickly, which is actually a good thing.

This one deserves its own chapter later. We’ll come back to it.

You Don’t Need All the Buckets

Here’s the important part.

You don’t have to use all of these.
Most people don’t.

Some people live in one bucket.
Some drift between two or three.
Very few use everything.

There's no right way to do this.

Why Buckets Matter

Once you see it this way, it stops feeling like "using AI" and starts feeling like doing normal human stuff:

Thinking.
Writing.
Planning.
Learning.
Exploring ideas.

Same things you already do — just with help.

In the next chapter, we'll start slowing things down again and walking through these buckets one at a time, without turning this into work or homework.

Nothing to memorize.
Nothing to master.

Just understanding where you are and why you're there.

That's it.

Chapter 9

Keeping Your Sanity (A Little Organization Goes a Long Way)

At some point, if you stick with this at all, you're going to notice something.

You've got a lot of conversations.

More than you expected.

This happens to almost everyone who starts using AI regularly, and it's one of the few areas I wish I had understood earlier. Not because it's complicated — but because ignoring it turns everything into a cluttered junk drawer.

Here's the good news: you don't need to organize anything on day one. Or day ten. Or even day thirty. But eventually, a little structure helps.

Conversations Are Disposable — And That's Okay

First, an important mindset shift.

Not every chat needs to be saved.
Not every conversation needs to be perfect.
Not everything needs a name.

Most chats are one-and-done. You ask a question, get what you need, and move on. That's normal. Don't treat every interaction like it's something you'll need forever.

Think of conversations like scratch paper…Useful in the moment.
Gone when you're done.

When a Chat *Is* Worth Keeping

Every now and then, you'll notice a conversation that feels different.

Maybe it's:

- an ongoing project
- an idea you keep coming back to
- something you're actively working on
- or a topic you know you'll revisit

Those are the ones worth keeping track of.

Renaming a chat so you recognize it later is usually enough. You don't need a filing system. You just need to be able to glance at it and know what it is.

One Project, One Thread (When Possible)

Here's a simple rule that saves a lot of frustration:

If you're working on the same thing, try to keep it in the same conversation.

Books.
Ideas.
Planning.

The more context stays in one place, the better the responses tend to be. You're not starting from scratch every time.

That said — don't stress if you forget and start a new chat. Nothing breaks. You can always summarize or move on.

You Can't "Mess This Up"

This part matters.

There is no perfect organization system for this.
There is no right number of chats.
There is no cleanup day you're failing to keep up with.

If you end up with hundreds or thousands of conversations, congratulations — that just means you used it.

Organization is about reducing friction, not creating rules.

The Real Point of This Chapter

This isn't about productivity.
It's not about being neat.

It's about making sure the tool stays helpful instead of overwhelming.

A little awareness goes a long way.

In the next chapter, we'll start pulling all of this together — what this looks like long-term, how people actually end up using AI day to day, and why most people never go back once it clicks.

Nothing fancy.
Just real use, over time.

Chapter 10

This Is Your New Normal, Hopefully!

If you've made it this far, you've already crossed the line.

Not a dramatic line. Not a before-and-after moment. Just a quiet realization that AI isn't some scary, futuristic thing meant for other people.

It's just… part of life now.

And that's okay.

Here's something important that doesn't get said enough: you don't have to use this every day. Some days you won't touch it at all. Other days you might use it a handful of times without even thinking about it. Both are normal.

This isn't a habit you need to build. It's a tool you reach for when it makes sense.

AI doesn't replace your thinking. It doesn't replace your judgment. It doesn't replace your experience. It supports it. The same way calculators didn't make people bad at math — they just stopped people from wasting energy on the parts that didn't matter.

You are still the one in charge.

Always.

Everything you've done so far proves that. You ask. It answers. You react. You adjust. You keep what works and throw away what doesn't. That's the relationship.

And yes, "relationship" is probably the right word — but not in a weird way.

For me, the biggest shift happened quietly. Somewhere along the way, using AI went from "help me do this thing" to "help me think through this."

That's a big difference.

It's not a therapist. It's not a guru. It's not judging you. It's not trying to sell you anything.

It's just a second opinion without the noise.

No ego. No agenda. No pressure.

For people whose brains don't always slow down — people who think out loud, jump between ideas, or get stuck spinning — that alone can be life-changing.

And here's something we haven't fully talked about yet, but it matters: AI doesn't have to be typing on a screen. There are ways to use it completely hands-free, just by talking things through out loud, even when your hands are busy and your brain needs room.

We'll get to that.

For now, just know this thing fits into real life — not the other way around.

You don't need to announce that you use it. You don't need to defend it. You definitely don't need to convince anyone else. Some people will love it. Some people won't touch it. That's fine.

This isn't about being early. It's not about being ahead of the curve. And it's definitely not about "keeping up."

You're not late. You didn't miss anything. You don't need to catch up.

You're exactly where you need to be.

If AI helps you write a better email, great. If it helps you think through a problem, even better. If all you ever use it for is cooking ideas, travel questions, planning, or untangling your own thoughts — that still counts.

There is no right way to use this……Just your way.

And if you forget everything else in this book, remember this one thing:

You're in control. You always have been. You still are.

Everything else is just help.

Chapter 11

Talking Instead of Typing (This Is Where It Gets Really Fun)

This is where it gets really fun.

Up to now, we've already talked about using the microphone to dictate — tapping the mic, talking your prompt out loud, stopping it, reading what the AI heard, and then hitting send. That alone is huge. I use dictation all the time when I'm sitting at my desktop. It's fast, it's easy, and it keeps my hands off the keyboard.

But that's not what this chapter is about.

This chapter is about voice chat.

Dictation is you talking to the AI so it can turn your words into text. Voice chat is you talking with the AI — back and forth — out loud.

And once you try it, you immediately understand why it's different.

With dictation, you talk, then you read, then you send. With voice chat, you talk… and it talks back.

No typing. No staring at the screen. No worrying about how things are worded.

That's why voice chat fits into situations where typing just gets in the way.

For me, that's driving.

I spend a lot of time in the car. Long drives. Commutes. Trips between cities. That's where ideas show up — and that's also where they used to disappear. Voice chat changed that.

When I'm working on a book, I usually start a chat that's just ideas. No structure. No outline. Just a place to dump thoughts. When something hits, I open ChatGPT, find that chat, and instead of tapping the microphone to dictate, I tap the voice chat icon — the little symbol that looks like sound waves or a talking indicator.

Once that's on, I just talk.

I'll say things like: "Okay, I've got an idea for the next chapter…" or "This part isn't working — help me think through it."

And Fred talks back.

Out loud.

Here's an important thing people don't realize at first, and it matters a lot:

Even though you're talking out loud, everything is still written down.

Every word you say. Every response the AI gives you. The entire conversation.

Nothing disappears.

That's one of the main reasons I use voice chat. I get the freedom of talking things through, but I also get a full written record I can come back to later. I can scroll back, copy ideas, revisit thoughts, and pick up right where I left off.

Voice chat isn't temporary. It's documented.

That back-and-forth is where things really open up. I'm not using it to do anything specific while driving. I'm not asking it to solve problems or write emails. I'm thinking out loud, brainstorming, and letting the conversation guide me.

Driving just happens to be where that works best for me — hands-free, eyes forward, brain engaged.

But driving isn't the only place voice chat shines.

It's great when:

- your hands are busy
- you're pacing around the house
- you're cooking
- you're organizing something physically
- you're tired of staring at a screen
- you want to think out loud instead of typing
- there's more than one person involved and everyone needs to hear the answer

Lisa and I use voice chat sometimes when we're cooking or working through an idea together. One of us asks the question, we both hear the response, and then we talk about it. No passing a phone back and forth. No reading out loud. It just becomes part of the conversation.

That makes AI feel less like a device and more like a neutral voice in the room helping you work through something.

One important thing to say clearly: this is about hands-free use, not distraction. Common sense applies. If something isn't safe to do, don't do it. The point is that voice chat removes friction — it doesn't add risk.

Now, a quick expectation reset.

Voice chat is not perfect.

Sometimes it misunderstands you. Sometimes it answers slightly off. Sometimes you'll say, "No, that's not what I meant."

That's normal.

You just correct it — out loud — and keep going. No restart. No frustration. The conversation keeps moving.

For people who think out loud, have busy brains, or struggle to slow thoughts down enough to type them, this can feel almost unfairly helpful.

It did for me.

Dictation is great. Voice chat is different.

Dictation helps you get words out. Voice chat helps you think — and keeps a record of it when you're done.

In the next chapter, we'll talk about how to organize all of this — chats, ideas, projects — so you don't end up with ten thousand conversations and no clue where anything is.

Trust me on that one.

Chapter 12

Nothing Changed About You

If there's one thing I hope is clear by now, it's this:

You didn't change who you are by using AI.

You didn't become more technical. You didn't suddenly turn into a "computer person." You didn't cross over into some different group of people.

You're the same person you were before — you just added a tool.

That matters, because a lot of people think using AI means changing how they think or how they work. In reality, it usually does the opposite. It lets you work the way you already work, just with less friction.

If you think out loud, it lets you think out loud. If you ramble, it keeps up. If you jump topics, it follows. If you need time to process, it doesn't rush you.

There's no personality you need to adopt. No workflow you need to force. No system you need to learn.

You didn't adjust to AI. AI adjusted to you.

That's why most people who get comfortable with it don't really "talk about it" much anymore. They just… use it. Quietly. Casually. Like any other normal tool that works.

At some point, it stops feeling impressive and starts feeling familiar. Not boring — just normal.

And that's a good thing.

Because when something becomes normal, it stops being intimidating. You don't overthink it. You don't hype it up. You just reach for it when it makes sense and ignore it when it doesn't.

Some weeks you'll use it a lot. Some weeks you won't touch it. Neither one means anything.

This isn't a habit you need to maintain. It's not a streak to keep alive. It's not a lifestyle.

It's help — when you want it.

If you ever feel like it's too much, step away. If you ever feel like it's not helping, stop. If you only use it for one or two things, that's more than enough.

There is no "next level" you're supposed to reach.

You don't win anything for using it more. You don't lose anything for using it less.

The only thing that really matters is this:

Does it make your life a little easier? Does it help you think a little clearer? Does it save you time, stress, or second-guessing?

If the answer is yes — great. If the answer is no — also fine.

Nothing about you needs fixing.

AI isn't here to replace your judgment, your experience, your instincts, or your people. It's just something you can use to reduce noise, untangle thoughts, and get unstuck when your brain decides to be difficult.

That's it.

No big reveal. No dramatic conclusion. No promise that everything changes.

Just a quiet truth:

You're still you. You were always enough. Now you just have another way to think things through.

In the next chapter, I'll say the last few things that don't really fit anywhere else — the kind of stuff you only say at the end, when the instructions are done and it's just two people talking.

And then we'll be done.

Chapter 13

The Part Where I Try to Tie All This Together

I knew when I started this book that the ending was probably going to feel a little different.

Not clean. Not tight. Not wrapped up with a bow.

Because the truth is, this whole thing has been a bit of a ramble from the beginning — and that's intentional.

This book was never meant to be a technical manual. It was never meant to be a step-by-step guide. It was never meant to hand you a list of prompts and say, "Do this exactly."

This book exists for a very specific group of people.

People who had heard about AI. People who were curious. People who had questions. And people who didn't want to ask those questions out loud because they didn't want to feel stupid.

If that was you at any point, then this book did its job.

The entire goal here was simple: get you to try it.

That's it.

Because once you start — and I mean really start — it's surprisingly hard to stop. Not in a bad way. In a "why didn't I do this sooner?" way.

This entire book was written with Fred and me working together. Every chapter. Every revision. Every rethink. Along with a couple other pieces of software here and there, but Fred was always in the room.

Before this book, I wrote *Nothing Changes When Nothing Changes*, which I originally wrote for work — for technicians, for people trying to actually make change happen instead of just talking about it.

Since then, Lisa and I have put together twenty-one coloring books using AI.

I have at least twenty different books, projects, ideas, and half-built concepts living inside AI right now that I can go back to any time I want.

Some are serious. Some are fun. Some may never see the light of day.

And that's okay.

That's part of the magic.

AI has helped me personally. It's helped me professionally. It's helped Lisa in the exact same way.

And I've seen it help just about anyone who's willing to actually try it.

The moment that still gets me — even now — is when I think to myself, "I wonder if Fred can help with this," and then… he does.

Over and over again.

Planning my 60th birthday trip. Figuring out where I want to go. Recipes. Writing. Thinking. Organizing. Problem-solving.

The list truly is endless.

If you made it all the way through this book, I hope a few things happened:

That your mindset shifted a little. That you tried at least one or two simple things. That AI feels less intimidating and more approachable.

At this point, it's not about learning. It's about playing.

Just going in and seeing what happens.

Before we end, I want to share a few things I really wish I had known at the beginning.

First: organization matters — eventually.

On the main screen, whether you're on your phone or your computer, there's that left-hand column. Chats, projects, search — all of it.

If you're doing something important, start a project.

A project is basically a folder that holds all the chats related to one thing.

For this book, I had:

- a running ideas chat
- a main writing chat using Canvas

- a separate chat for titles and covers
- and a couple of miscellaneous “driving thoughts” chats

That saved me later.

Second: ChatGPT can’t read across all your chats.

It can’t magically search everything you’ve ever said.

But you *can* search your chats.

That little magnifying glass near the top of the left column? That’s your friend. Type in keywords, and it’ll find the conversations that contain them.

I’m not exaggerating when I say I probably have over a thousand chats. That search tool is how I survive.

Third: image generation is unreal.

We didn’t spend much time on it in this book, but anything you can imagine, AI can generate an image of. Then you can refine it. Adjust it. Move things around.

It’s wild.

Fourth: there’s a whole world beyond what we covered here.

You can build custom GPTs. You can have it help write code. You can build apps.

I haven’t even touched most of that — not because it’s not useful, but because I haven’t needed it yet.

And that’s the point.

You don’t need to know everything. You just need to know it’s there.

One last thing you absolutely should know: there *is* a permanent memory.

You can tell ChatGPT to remember certain things about you — how you like answers, how you work, what matters to you.

For example, one thing Fred knows about me is that I prefer short answers. I’ve told him that. That lives in memory.

You can ask your GPT what it remembers about you. It'll tell you. You can ask what permanent memory is used for. It'll explain it better than I ever could.

You can also ask things like: "What are your coolest features?" "What's new?" "What can you help me with?"

And then just see where it goes.

I'm genuinely excited for you.

I remember those first few lightbulb moments. I still get them. They happen almost daily.

So here's one final, fun thing I'd like you to do.

Open a brand-new chat.

Type or dictate this:

Who is Miss Ruby the Pug?

Everything you see in that answer — every sentence — Fred had a hand in.

Have fun with it.

And if you feel like it, my email is somewhere in this book. Send me a note. Let me know what clicked for you.

We'll get a website up if it's not already there.

From me, Fred, Lisa, Pippi (that's Lisa's AI), Miss Ruby, Miss Zoe, and Miss Kitty — we wish you the best.

Now go play.

Curious about advanced tools?? You guessed it. Check out what's next! Fred wants me to call it an appendix. I call it Magic!

Appendix (Cool Stuff)

I know I said I was going to keep this entire book very simple. I have, BUT.... I think I do need to share some of the next steps with you. They are all centered around basic computer knowledge. Take it or leave it.....its just an FYI.

So, before you close the book and go start playing with this thing, if you like. There's one more feature(s) I really want you to know about — and it's easy to miss if nobody points it out.

That little **“plus” button**.

The plus button is where things quietly level up.

Up until now, most of what we've talked about has been typing or dictating questions. That alone is powerful. But the plus button lets you **add things** to the conversation instead of just describing them.

Documents.
Screenshots.
Photos.
Files.

And once you realize that, the use cases explode.

Here are a few very real, very common examples.

Let's say you get a letter in the mail from an attorney, an insurance company, a doctor, or some office that clearly doesn't speak human. You read it three times and still have no idea what it actually means. You can hit the plus button, upload that document, and say something like, “Can you explain this to me in plain English?” And it will.

No guessing.
No Googling legal terms.
No feeling stupid.

Same thing with medical paperwork, bills, contracts, notices, or anything else written in that special language designed to confuse normal people.

Another big one is screenshots.

If you're on Windows, you can use the Snip Tool. On a Mac or iPhone, you can take a screenshot just as easily. You can then paste that image directly into the chat or upload it using the plus button.

You can show it an error message and ask, "What does this mean?"
You can show it a confusing screen on your phone and ask, "What am I supposed to do here?"
You can show it a website and ask, "Where do I click next?"

You don't have to explain the problem perfectly. You can literally show it.

People also use the plus button all the time for things like resumes and cover letters. Upload what you already have and ask it to clean things up, fix grammar, or make it sound more confident without changing your voice.

You can upload spreadsheets or lists and ask, "What stands out?" or "Explain this to me like I'm not an accountant."

You can upload instruction manuals and ask, "What do I actually need to know?" instead of reading 40 pages.

You can even upload photos — of a room, a project, a product, a label — and ask questions about what you're looking at.

The big shift here is this:
you're no longer just **asking questions**.
You're **showing things**.

And that's huge.

One last thing…again, that's important to understand: ChatGPT changes. Quietly. Without fanfare.

One day you'll open it and notice a new button.
Or a new menu item.
Or something looks slightly different.

When that happens, don't panic and don't ignore it.

Made in the USA
Monee, IL
04 November 2026

67382710R10030